Sukaina Hakami (Ed.)

**Vertical Handoff in Fourth Generation (4G) Wireless Networks**

Sukaina Hakami (Ed.)

# Vertical Handoff in Fourth Generation (4G) Wireless Networks

LAP LAMBERT Academic Publishing

**Impressum / Imprint**

Bibliografische Information der Deutschen Nationalbibliothek: Die Deutsche Nationalbibliothek verzeichnet diese Publikation in der Deutschen Nationalbibliografie; detaillierte bibliografische Daten sind im Internet über http://dnb.d-nb.de abrufbar.

Alle in diesem Buch genannten Marken und Produktnamen unterliegen warenzeichen-, marken- oder patentrechtlichem Schutz bzw. sind Warenzeichen oder eingetragene Warenzeichen der jeweiligen Inhaber. Die Wiedergabe von Marken, Produktnamen, Gebrauchsnamen, Handelsnamen, Warenbezeichnungen u.s.w. in diesem Werk berechtigt auch ohne besondere Kennzeichnung nicht zu der Annahme, dass solche Namen im Sinne der Warenzeichen- und Markenschutzgesetzgebung als frei zu betrachten wären und daher von jedermann benutzt werden dürften.

Bibliographic information published by the Deutsche Nationalbibliothek: The Deutsche Nationalbibliothek lists this publication in the Deutsche Nationalbibliografie; detailed bibliographic data are available in the Internet at http://dnb.d-nb.de.

Any brand names and product names mentioned in this book are subject to trademark, brand or patent protection and are trademarks or registered trademarks of their respective holders. The use of brand names, product names, common names, trade names, product descriptions etc. even without a particular marking in this work is in no way to be construed to mean that such names may be regarded as unrestricted in respect of trademark and brand protection legislation and could thus be used by anyone.

Coverbild / Cover image: www.ingimage.com

Verlag / Publisher:
LAP LAMBERT Academic Publishing
ist ein Imprint der / is a trademark of
OmniScriptum GmbH & Co. KG
Heinrich-Böcking-Str. 6-8, 66121 Saarbrücken, Deutschland / Germany
Email: info@lap-publishing.com

Herstellung: siehe letzte Seite /
Printed at: see last page
**ISBN: 978-3-659-75920-8**

Zugl. / Approved by: Riyadh, Princess Nourah University, 2015

Copyright © 2015 OmniScriptum GmbH & Co. KG
Alle Rechte vorbehalten. / All rights reserved. Saarbrücken 2015

# Vertical handoff in Fourth Generation (4G) Wireless Networks

By

Sukaina Hussain Hakami

Sukaina.alh@gmail.com

Amjad Saad BinSlmah

a.s.binslmah@gmail.com

Jawaher Dhafer Alshahrani

ms-joje@hotmail.com

Lojain Ibrahim AlAfandi

lojain.i.a@hotmail.com

Maram Ghorm Al-Shehri

Supervised BY

Dr. Samia Chelloug
College of Computer Sciences and Information Sciences
in Partial Fulfillment of the Requirements for the
Degree of Bachelor of Science
in Networking And Communication System

Riyadh, KSA

1435 - 1436H

# Acknowledgments

First we would like to thank Allah for helping us through all the years to reach this level of knowledge. Also we extend our sincere thanks and appreciation to our supervisor Dr. Samia Chelloug for being our monitor and guide, helping us complete our project with her continues support to improve our research and teamwork skills.

Finally, we would never achieve all of this if it wasn't our parents. We deeply thank them for all their long nights taking care of us and being always there for us whenever we needed them. They've been always supporting us to chase our dreams and to be better individual.

# Table of content:

Acknowledgments ..................................................................................................... ii

List of Figures ......................................................................................................... vii

List of Abbreviations ............................................................................................. viii

Abstract .................................................................................................................... ix

Chapter 1: Introduction ............................................................................................. 1

    1.1 Problem statement and significance ................................................................ 2

    1.2 Proposed solution ............................................................................................ 2

    1.3 Project domain and Limitation ........................................................................ 2

    1.4 Definition of new terms .................................................................................. 3

Chapter 2: Background information and related works ........................................... 5

    2.1 Fourth generation ............................................................................................ 6

        2.1.1 Introduction: ........................................................................................... 6

        2.1.2 First generation: ..................................................................................... 6

        2.1.3 Second generation: ................................................................................ 6

        2.1.4 2.5 generation: ....................................................................................... 7

        2.1.5 Third generation: ................................................................................... 7

        2.1.6 Fourth generation: ................................................................................. 8

    2.2 Handoff: ......................................................................................................... 11

        2.2.1 Introduction: ......................................................................................... 11

        2.2.2 Handoff definition: .............................................................................. 11

        2.2.3 Types of handoffs: ............................................................................... 11

        2.2.4 VHO Process: ...................................................................................... 12

        2.2.5 Classification of VHO: ........................................................................ 13

        2.2.6 Mobile Controlled and Network Controlled handoffs: ...................... 14

        2.2.7 Issues: ................................................................................................... 14

        2.2.8 Vertical Handover Decision (VHD) Algorithms: ............................... 15

- 2.3 Localization Methods ..................................................................... 16
  - 2.3.1 Angle of Arrival .................................................................... 16
  - 2.3.2 Signal strength ..................................................................... 16
  - 2.3.3 Uplink Time of Arrival ........................................................ 17
  - 2.3.4 Downlink Observed Difference ........................................... 17
  - 2.3.5 Location Pattern Matching .................................................. 18
- 2.4 Related work and survey ............................................................... 18
  - 2.4.1 Vertical handover decision algorithm: ................................. 18
  - 2.4.2 Vertical Handoff Scheme between Mobile WiMax and Cellular Networks Based on the Loosely Integration Model: ................................. 21
  - 2.4.3 Conclusion: .......................................................................... 21

Chapter 3: System analysis ....................................................................... 22
- 3.1 Requirements Specification .......................................................... 23
- 3.2 Requirements analysis .................................................................. 23
  - 3.2.1 Results Analysis: ................................................................. 23

Chapter 4: System design ......................................................................... 26
- 4.1 System Architecture ..................................................................... 27
- 4.2 Comparison between our system and related works ..................... 30
- 4.3 User interface design .................................................................... 31
  - 4.3.1 Modeling Concepts: ............................................................. 31
  - 4.3.2 Interface simulation ............................................................. 33

Chapter 5: Implementation ....................................................................... 34
- 5.1 Implementation Requirement ....................................................... 35
  - 5.1.1 Hardware: ............................................................................ 35
  - 5.1.2 Software: ............................................................................. 35
  - 5.1.2.3 Modules and protocols: .................................................... 36
  - 5.1.2.4 Communication between protocol layers ......................... 37

5.2 Implementation Details: ................................................................................. 39

    5.2.1 Simulation parameters: ....................................................................... 46

    5.3 I/O screens: ............................................................................................ 47

Chapter 6: Testing ............................................................................................. 52

    6.1 Test plan: ................................................................................................ 53

    6.2 Test case: ................................................................................................ 53

Chapter 7: Conclusion ....................................................................................... 56

    7.1 Evaluation: ............................................................................................. 57

    7.2 Future work: .......................................................................................... 57

References: ........................................................................................................ 58

Appendices: ....................................................................................................... 60

## List of Tables:

Table 1: Difference between 3G and 4G...................................................9
Table 2: Comparison between our system and related works................30
Table 3: Description of Mobility Architecture........................................41
Table 4: Description of StandardHost......................................................42
Table 5: Description of ITCPQ Architecture..........................................43
Table 6: Description of IUDP Architecture ............................................43
Table 7: Description of Network Layer Architecture .............................44
Table 8: Description of Loopback interface Architecture.......................45
Table 9: Description of Iexternalnic Architecture...................................45
Table 10: Simulation parameters..............................................................46
Table 11: Test case....................................................................................53

# List of Figures

Figure 1: Handoff Types ..................................................................................12
Figure 2: Upward and Downward Handoffs ..............................................13
Figure 3: Hard and Soft Handovers ............................................................13
Figure 4: Mohanty and Akyildiz's VHD heuristic ....................................19
Figure 5: Flowchart for our system ............................................................27
Figure 6: Bounding box algorithm ..............................................................30
Figure 7: Interface Simulation ....................................................................33
Figure 8: Mobility Architecture ..................................................................41
Figure 9: ITCP Architecture .........................................................................42
Figure 10: IUDP Architecture ......................................................................43
Figure 11: Network Layer Architecture ....................................................44
Figure 12: Loopback interface Architecture ............................................44
Figure 13: IExternalNic Architecture .........................................................45
Figure 14: IWiredNic Architecture ..............................................................45
Figure 15: RSSI average on 200 speed result ...........................................53
Figure 16: Probability on 200 speed result ...............................................54
Figure 17: RSSI on multiple speeds result ................................................54
Figure 18: RSSI on 200 speeds result ..........................................................55

**List of Abbreviations**

| 3G | Third Generation |
|------|------------------|
| 4G | Fourth Generation |
| WLAN | Wireless Local Area Network |
| RSSI | Received Signal Strength Indicator |
| VHD | Vertical Handover Decision |

## Abstract

In this project we focused on the heterogeneity in the 4G (fourth generation) which allows a high speed and capacity. The main idea consists to track a mobile user moving from a WLAN (Wireless Local Area Network) to a 3G (third generation) network.

The decision of the vertical handoff is based on the strength of the signal. In other hand, the tracking process is performed using bounding box. More specifically, the handoff execution phase starts once the user is tracked at or near the border of a cell.

Our results are obtained through OMNeT++ and show the impact of the distance and the speed on the signals strength and the handoff probability.

**Chapter 1: Introduction**

## 1.1 Problem statement and significance

Handoff is one of the issues of 4G communication. It requires resources (power and time) and it causes unwanted interruption to the connection between different networks of WLAN toward 3G, occurs either by crossing the cell's borders or signal weakness. The users want to be all along connected to a base station so it is considered as a problem for the users to be disconnected for a while till they connect to the other base stations. There are many ways for the handoff detection some of them are not accurate and other are so complicated or consume much power. In our project we will propose a simple and accurate method to track a mobile user and take handoff decision in the appropriate time.

## 1.2 Proposed solution

We assumed that we have two different networks one is a WLAN and the other one is operating under 3G. The communication between them is half duplex and the user movement will be in a vertical way. By using the Bounding Box algorithm the user will be tracked once the power of the signal is below a certain threshold (To make sure that the user is ready to move to another base station) the handoff may starts if the user decide to move.

## 1.3 Project domain and Limitation

4G network is composed of different wireless sub networks that complement each other and like any other mobile network in dealing with handoffs and it got two types of them the horizontal handoff between two base stations in the same system and the other type is the vertical handoff between two or more wireless networks in a manner that is completely transparent to applications and disrupts connectivity as little as possible it has two categories, upward vertical handoff & downward vertical handoff. 4G has been improved in the wireless side so the advanced wireless networks 4G communication is to be able to handle much higher data rates in WLAN and cellular networks. A user with a large range of mobility will access the network and will be able to seamlessly reconnect to different networks, even within the same session. The spectra allocation is expected to be more flexible and even flexible spectra sharing among the different sub networks are anticipated. There are challenges in the design and these are the primary ones: Low Latency Handoff, Power Savings, Bandwidth Overhead, Discover the right time to perform handoffs in a wireless channel that is difficult to predict and characterize.

## 1.4 Definition of new terms

- **Handoff definition:**

    Handoff is the process of changing the channel (frequency, time slot, spreading code, or combination of them) associated with the current connection while a call is in progress. It is often initiated either by crossing a cell boundary or by deterioration in quality of the signal in the current channel.

- **Vertical handoff:**

    It allows a mobile user to move between two or more wireless networks in a manner that is completely transparent to applications and disrupts connectivity as little as possible.

- **Received signal strength (RSS):**

    The received signal strength is a function of distance between the transmitter and receiving device, which varies due to various in-path interferences. The RSS measurements from distant beacons contain useful information that can improve the quality of localization algorithms despite the high degree of uncertainty. [12]

- **Bounding Box**

    The bounding box algorithm is a computationally simple method of localizing nodes given their ranges to several beacons.[11]

- **Long Term Evolution (LTE):**
    Is the last step toward the 4th generation of radio technologies designed to increase the capacity and speed of mobile telephone networks.

- **Worldwide Interoperability for Microwave Access (WIMAX):**

  Is a wireless telecommunication technology based on the IEEE 802.16 family of standards, which allows for high-speed wireless data transmission over long distances (5-30miles).[1]

# Chapter 2: Background information and related works

## 2.1 Fourth generation

### 2.1.1 Introduction:

Wireless emerging technologies offer a lot of opportunities because they are less expensive than wired technologies, may guarantee a good communication in a noisy environment, and allow users to move freely. More precisely, mobile generations are growing rapidly and aim to enhance the communication data rates. Our project focuses on the 4th generation, which supports network heterogeneity and represents the broadband mobile communications stage. It uses some techniques that are different from those used in the 3rd generation. So, this chapter explains the theoretical background of the fourth generation.

### 2.1.2 First generation:

1G refers to the first generation of wireless telephone technology. It was introduced in late 1980s but it was in use in ear ly 1990s. The speed was up to 2.4kbps. It was all about the voice calls. It uses analog signal.[2][4][5]

- **Drawbacks of 1G:**
    1. Poor Voice Quality
    2. Poor Battery Life
    3. Large Phone Size
    4. No Security
    5. Limited Capacity
    6. Poor Handoff Reliability

### 2.1.3 Second generation:

2G technology refers to the 2nd generation which is based on GSM. It was launched in Finland in the year 1991. 2G network uses digital signals. Its data speed was up to 64kbps.[4][5]

- **Features Includes:**
    1. It enables services such as text messages, picture messages and MMS (multimedia message).
    2. It provides better quality and capacity.

- **Drawbacks of 2G:**

    2G require strong digital signals to help mobile phones work. If there is no network coverage in any specific area, digital signals would weak. These systems are unable to handle complex data such as Videos.

### 2.1.4 2.5 generation:

2.5G is a technology between the second (2G) and third (3G) generation of mobile telephony. 2.5G is sometimes described as 2G Cellular Technology combined with GPRS. [4][5]

- **Features Includes:**
    1. Phone Calls
    2. Send/Receive E-mail Messages
    3. Web Browsing
    4. Speed: 64-144 kbps
    5. Camera Phones
    6. Take a time of 6-9 mins. To download a 3 mins. Mp3 song

### 2.1.5 Third generation:

3G technology refer to third generation which was introduced in year 2000s. Data Transmission speed increased from 144kbps- 2Mbps. Typically called Smart Phones and features increased its bandwidth and data transfer rates to accommodate web-based applications and audio and video files.[4]

- **Features Includes:**
    1. Providing Faster Communication
    2. Send/Receive Large Email Messages
    3. High Speed Web / More Security
    4. Video Conferencing / 3D Gaming
    5. TV Streaming/ Mobile TV/ Phone Calls
    6. Large Capacities and Broadband Capabilities

- **Drawbacks Of 3G:**

    1. Expensive fees for 3G Licenses Services.
    2. It was challenge to build the infrastructure for 3G.
    3. High Bandwidth Requirement.
    4. Expensive 3G Phones.
    5. Large Cell Phones.

And now we will talk about our main subject which is the fourth generation

### 2.1.6 Fourth generation:

4G technologies are sometimes referred to by the acronym "MAGIC," which stands for Mobile multimedia, Anytime/any-where, Global mobility support, integrated wireless and Customized personal service.[1][2]

- **Technology Used:**

    1. TDMA: Time Division Multiple Access, is a technique for dividing the time domain up into sub channels for use by multiple devices.
    2. CDMA: Code Division Multiple Access, allows every device in a cell to transmit over the entire bandwidth at all times. [3]

- **4G Hardware:**

    1. Ultra Wide Band Networks: Ultra Wideband technology, or UWB, is an advanced transmission technology that can be used in the implementation of a 4G network.
    2. Smart Antennas: Multiple "smart antennas" can be employed to help find, tune, and turn up signal information.[3]

- **General 4G Services and 4G Applications:**

    1. Localized/Personalized Information.
    2. Organizational services.
    3. Communications services and applications
    4. Entertainment services
    5. Mobile commerce (M-Commerce )

- **Features of 4G Wireless Systems:**
    1. Support interactive multimedia.
    2. User friendliness.
    3. High speed, high capacity and low cost per bit.
    4. Higher band widths.
    5. Terminal Heterogeneity.
    6. Network Heterogeneity.

- **User and Industry Expectations:**
    Wireless users can be categorized into generalized segments:
    1. The Age segment
    2. The Internet Usage segment
    3. The Mobile Professional segment

- **Difference between 3G and 4G:**

Table 1: Difference between 3G and 4G

| Major Requirement | 3G | 4G |
| --- | --- | --- |
| Speed | 384 Kbps to 2 Mbps | 20 to 100 Mbps |
| Frequency Band | Dependent on country | Higher frequency bands (2-8 GHz) |
| Bandwidth | 5-20 MHz | 100 MHz (or more) |
| Switching Design Basis | Circuit and Packet | All digital with packetized voice |
| Access Technologies | W-CDMA | TDMA, CDMA |

**Types of 4G:**

Historically, WiMAX and Long-Term Evolution (LTE), the standard generally accepted to succeed both CDMA2000 and GSM, have been marketed and labeled as "4G technologies," but that's only partially true: they both make use of a newer, extremely efficient multiplexing scheme (OFDMA, as opposed to the older CDMA or TDMA), however, WiMAX tops at around 40Mbps and LTE at around 100Mbps theoretical speed. Practical, real world commercial networks using WiMAX and LTE range between 4Mbps and 30Mbps. Even though the speed of WiMAX and LTE is well short of IMT-Advanced's standard, they're very different than 3G networks and carriers around the world refer to them as "4G". Updates to these standards WiMAX 2 and LTE-Advanced, respectively will increase throughput further, but neither has been finalized yet.[1]

- **Long Term Evolution (LET):**

  Is the last step toward the 4th generation of radio technologies designed to increase the capacity and speed of mobile telephone networks where the current generation of mobile telecommunication networks are collectively known as 3G (for "third generation"), LTE is marketed as and called 4G insinuating that it's the "fourth generation".

  Most major mobile carriers are converting their networks to LTE since 2009. LTE is a set of enhancements to the Universal Mobile Telecommunications System (UMTS) introduced in 3rd Generation Partnership Project (3GPP) Release 8.

  Much of 3GPP Release 8 focuses on adopting 4G mobile communications technology, including an all IP flat networking architecture.[1]

- **Worldwide Interoperability for Microwave Access (WiMAX):**

  Is a wireless telecommunication technology based on the IEEE 802.16 family of standards, which allows for high-speed wireless data transmission over long distances (5-30miles).

  The initial version, based on 802.16a, is designed for fixed (non-mobile) applications only, such as a wireless replacement for home DSL or cable modem service. Newer versions, such as 802.16e, add support for mobility, potentially making WiMax a competitor for certain 3G or 4G cell-phone technologies.

  WiMax operates at higher frequencies than mobile phone networks. WiMax technology can operate in the 2.5 or 3.5 GHz licensed bands, or in the 5.8 GHz unlicensed band. The name WiMAX was created by the

WiMAX Forum, which was formed in June 2001 to promote conformance and interoperability of the standard.[1]

## 2.2 Handoff:

### 2.2.1 Introduction:

The handoff is the process of establishing a new connection while the user is moving. Two types of handoff may occur in the $4^{th}$ generation: vertical and horizontal. Since VHO (vertical handoff) is an asymmetric process, the MT (Mobile Terminal) moves between two different networks with different characteristics. So, it is necessary to select the best network, which provides high performance. The VHO operation should provide a minimum overhead, authentication of the mobile users and the connection should be maintained to minimize the packet loss and transfer delay.

### 2.2.2 Handoff definition:

Handoff is the process of changing the channel (frequency, time slot, spreading code, or combination of them) associated with the current connection while a call is in progress. It is often initiated either by crossing a cell boundary or by deterioration in quality of the signal in the current channel.

### 2.2.3 Types of handoffs:

We have two types of handoffs in 4G networks

- **Horizontal handoff:**
  Handoff between two base stations of the same system. Horizontal handoff involves a terminal device to change cells within the same type of network to maintain service continuity.

- **Vertical handoff:**

    It's allows a mobile user to move between two or more wireless networks in a manner that is completely transparent to applications and disrupts connectivity as little as possible.

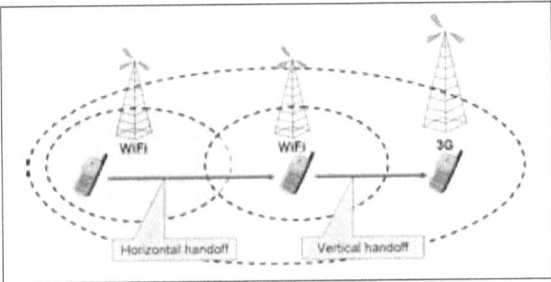

Figure 1: Handoff Types

### 2.2.4 VHO Process:

The vertical handoff process can be divided into three main steps namely handoff initiation, handoff decision, and handoff execution.

- **Handoff Initiation Phase:**

    In order to trigger the handoff event, information to be collected about the network from different layers likes Link Layer, Transport Layer and Application Layer. These layers provide the information such as RSS, bandwidth, link speed, throughput, jitter, cost, power, user preferences and network subscription etc. Based on this information handoff will be initiated in an appropriate time.

- **Handoff Decision Phase:**

    The mobile device decides whether the connection to be continued with current network or to be switched over to another one. The decision may depend on various parameters which have been collected during handoff initiation phase.

- **Handoff Execution Phase:**

    Existing connections need to be re-routed to the new network in a seamless manner. This phase also includes the authentication and authorization, and the transfer of user's context information.

### 2.2.5 Classification of VHO:

Vertical Handoff can be classified in to four types based up on its direction, process, control and decision:

### 2.2.5.1 Upward and Downward Handoffs:

In Vertical Handoff, if the mobile switches from the network with a small coverage to a network of larger coverage, it is termed as upward handoff. On the other hand, a downward handoff occurs in the reverse direction, i.e. from a network of larger coverage to a network of smaller coverage.

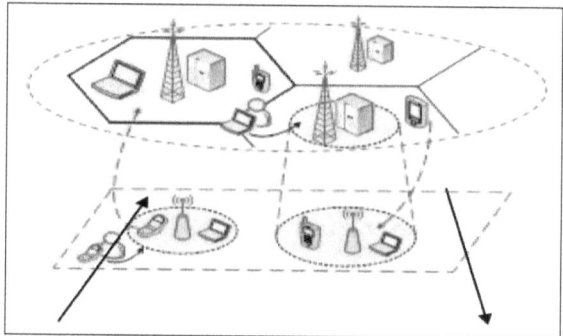

Figure 2: Upward and Downward Handoffs

### 2.2.5.2 Hard and Soft Handoffs:

When the mobile node switches to the target network only after the disconnection from current network is called as hard handoff or break before make. On the other hand, in soft handover a mobile node maintains the connection with the previous base station till its association with the new base station is completed. This process is also termed as make before break.

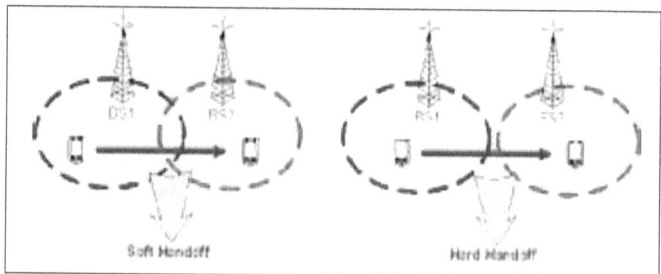

Figure 3: Hard and Soft Handovers

### 2.2.5.3 Imperative and Alternative handoffs:

When there is loss of signal strength an imperative handoff occurs. For imperative handoff the RSS is sufficient to be considered. On the other hand, an alternative vertical handoff is initiated to provide the user with better performance. For alternative handoffs several other network parameters such as available bandwidth, supported velocity and cost of the network are to be considered in addition to the device parameters such as quality of service demanded by the application and user preference.

### 2.2.6 Mobile Controlled and Network Controlled handoffs:

Vertical handoffs can further be classified based on who controls the handoff decision. If mobile node controls the handoff decision, it is termed as Mobile controlled handoff (MCHO). In Network controlled Handoff (NCHO) networks control the handoff decision. The handoff decision control is shared between the network and mobile in case of Mobile controlled Network Assisted (MCNA) and Network Controlled Mobile Assisted handoffs (NCMA). MCNA handoffs are more suitable because only mobile nodes have the knowledge about the network interfaces they are equipped with and user preferences can be taken into consideration.

### 2.2.7 Issues:

Future wireless systems will be based on heterogeneous wireless access technology. In order to provide seamless services many challenging issues to be solved.

- **QoS Issues**:
  Mobile terminals carrying real time and non-real time traffic should be serviced with guaranteed QoS. To provide best network service several parameters to be considered.

- **TCP Performance Issues:**
  When switching from low bandwidth, high data rate network to high bandwidth, low data rate network TCP performance should be considered for congestion.

- Security Issues:

Because of the wide coverage area when the sensitive data is transmitted it should be transferred in secured manner.

## 2.2.8 Vertical Handover Decision (VHD) Algorithms:

Several algorithms have been proposed in the research literature for use in the vertical handover decision (VHD) Brief description of each algorithm is as follows:

- Received signal strength (RSS)

Is the easiest way to measure the service quality and the most widely used criterion. RSS reading is directly related to the distance from the MT to its point of attachment. Most of the existing horizontal handover algorithms use RSS as the main decision criterion.

- Network connection time

Indicates the length of time that a user connected to an access point or base station, choosing the proper moment is very important to initiate a quality of service handover.

- Available bandwidth

Is a bit/sec expression that indicates the available data resources and is a measure of traffic conditions in the network. Signal to Interference and Noise Ratio SINR is related to available bandwidth algorithms.

- Power consumption

Refers to the MT's battery level, which becomes very important in case need to handover to another network that consumes lower power.

- Monetary cost

Some algorithms take into consideration the charging policies for different networks in making their handover decision.

- **Security**

  Integrity or confidentiality are considered as a critical issue in some applications, where the VHD may be chosen to a higher level of data security.

- **User preferences**

  Special user requirement or preference could be the issue that decides to initiate the handover.

## 2.3 Localization Methods

### 2.3.1 Angle of Arrival

Two APs measure the arrival angle of the signal, which is transmitted by a Device. The intersection area of the two lines determines the position of the Device. The main drawback of this technique is that a line of sight is required. Directional antennas are needed.

- **Advantages:**
    1. Only 2 A.
    2. Good accuracy under good propagation conditions.

- **Drawbacks:**
    1. LoS needed.
    2. Narrowing of the angle.
    3. Large number of users.
    4. Expensive (directional antennas).

### 2.3.2 Signal strength

The Device measures the strength of the AP signal and sends it back to the AP: indeed, in Bluetooth frame there is a possibility to get the Ratio Signal Strength Indicator (RSSI) calculated by the Device. Thus, because the power received is proportional to the distance between two Devices, when three APs get this information, three circles can be obtained and the intersection of them defines the probable location of the Device. The main drawbacks are the need of at least three APs and preferably an environment with LoS.

- **Advantages:**
    1. Low-cost.
    2. Easy to implement because no add software.
    3. Good accuracy under good propagation conditions.
- **Drawbacks:**
    1. High number of APs.
    2. Accuracy depends on the propagation channel.

### 2.3.3 Uplink Time of Arrival

Time of Arrival: APs measure the time for the signal to arrive from the Device. Because this measurement is directly related to the distance between the two stations, triangulation method can be used. Hyperbolas are obtained and their intersections give the location of the Device. Timing Differences technique: If stations are not synchronized, TDoA is used to determine the relative time of arrival between the 2 stations. The main drawback is that this method needs 3 APs.

- **Advantages:**
    1. Relative accuracy.
    2. No changes on Devices.
- **Drawbacks:**
    1. High number of APs.
    2. Large number of users.
    3. No location while Device is idle.

### 2.3.4 Downlink Observed Difference

Downlink Observed Difference is a Timing Differences technique. Measurements are made by the Device, which measures the time difference of the signals from several APs. Synchronization is needed between the Device and the APs.

- **Advantages:**
    1. Relative accuracy.
    2. Possible when Device is idle.
    3. Mobile-based implementation network possible.

- o **Drawbacks:**
    1- High number of APs.
    2- Software changes in Device.
    3- LoS needed if good accuracy wanted.

### 2.3.5 Location Pattern Matching

This technique compares the results obtained by measurements taken from a Device with some pre-trained sequences, simulated with some algorithms, on a server. These two techniques cannot be used in dynamic networks.

- o **Advantages:**
    1- Optimal accuracy in obstructed areas.
    2- Possible with one AP.

- o **Drawbacks:**
    1.Updates

## 2.4 Related work and survey

### 2.4.1 Vertical handover decision algorithm:
#### 2.4.1.1 RSS based VHD algorithms

RSS based VHD algorithms compare the RSS of the current point of attachment against the others to make handover decisions. Because of the simplicity of the hardware required for RSS measurements.

We described three representative RSS based VHD algorithms in the following sections and present a comparative summary of them in Table 1.

- o **An adaptive lifetime based handover heuristic**

    Zahran and Liang proposed an algorithm for handovers between 3G networks and WLANs by combining the RSS measurements either with an estimated lifetime metric (expected duration after which the mobile terminal will not be able to maintain its connection with the WLAN) or the available bandwidth of the WLAN candidate.

- **An RSS threshold based dynamic heuristic**

  Mohanty and Akyildiz proposed a WLAN to 3G handover decision method based on comparison of the current RSS and a dynamic RSS threshold (Sdth) when a mobile terminal is connected to a WLAN access point.

- **A traveling distance prediction based heuristic**

  Yan et al. developed a VHD algorithm that takes into consideration the time the mobile terminal is expected to spend within a WLAN cell.

  The method relies on the estimation of WLAN traveling time (i.e. time that the mobile terminal is expected to spend within the WLAN cell) and the calculation of a time threshold. A handover to a WLAN is triggered if the WLAN coverage is available and the estimated traveling time inside the WLAN cell is larger than the time threshold.

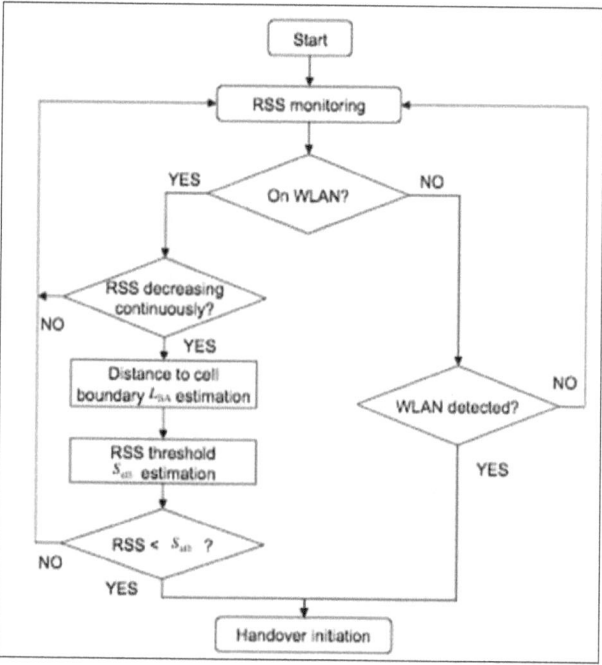

Figure 4: Mohanty and Akyildiz's VHD heuristic

### 2.4.1.2 Bandwidth based VHD Algorithms:

Bandwidth based VHD algorithms consider available bandwidth for a mobile terminal or traffic demand as the main criterion. In this section, three typical bandwidth based VHD algorithms are discussed in detail.

- **A QoS based heuristic**

    Lee et al. devised a QoS based VHD algorithm which takes residual bandwidth and user service requirements into account in deciding whether to handover from a WLAN to Wireless Wide Area Network (WWAN) and vice versa.

- **A signal to interference and noise ratio (SINR) based heuristic**

    Yang et al. presented a bandwidth based VHD method between WLANs and a Wideband Code Division Multiple Access (WCDMA) network using Signal to Interference and Noise Ratio (SINR).

### 2.4.1.3 Cost function based VHD algorithms

The cost function based algorithms combine metrics in a cost function. Many studies have been done in this area. In this section, we evaluate three representative cost function based VHD algorithms.

- **A multiservice based heuristic**

    Zhu and McNair's VHD algorithm relies on a cost function which calculates the "cost" of possible target networks.

    The algorithm prioritizes all the active applications, and then the cost of each possible target network for the service with the highest priority.

- **A cost function based heuristic with normalization and weights distribution**

    Similar to Zhu and McNair's method Hasswa et al. also proposed a cost function based handover decision algorithm in which the normalization and weights distribution methods are provided.

- **A weighted function based heuristic**

    Tawil et al. Presented a weighted function based VHD algorithm, which delegates the VHD calculation to the visited network instead of the mobile terminal.

### 2.4.1.4 Combination algorithms:

Combination algorithms are based on artificial neural networks or fuzzy logic, and combine various parameters in the handover decision such as the ones used in the cost function algorithm

### 2.4.2 Vertical Handoff Scheme between Mobile WiMax and Cellular Networks Based on the Loosely Integration Model:

Another algorithm for vertical handoff between Mobile WiMax and cellular networks was proposed which is based on the Loosely Integration Model. In this model, WLAN and 3G networks exist independently and provide autonomous services. For authentication and accounting for roaming services, a gateway was added to this incorporative model; for mobility between WLAN and 3G networks, this model also uses a mobile IP. One of the advantages of this model is that it can easily be adapted to existing communications and reduces the effort in developing new standards.

The Smoothly Integration Scheme algorithm has architecture similar to the Loosely Integration Model but only an IWG (Interworking Gateway) was added for interworking between Mobile WiMax and CDMA. The IWG helps by using an Extended Fast Handoff scheme in CDMA packets, which provides a gateway function for protocol adaptation. In the Fast Handoff scheme, the serving PDSN (Packet Data Serving Node) sends traffic to a target PDSN by setting up a tunnel. This traffic is forwarded to other mobile nodes by the target PDSN. In this method, the packet loss is minimized since the service anchor point is not changed

### 2.4.3 Conclusion:

After discussing some of the VHO Algorithms we got to the conclusion that every Algorithm got its steps and its own way to connect to the next network but they two got some issues that we analyzed in this paper. We also specialized in the Decision we explained the (VHD) and (VHSD) and their advantages.

# Chapter 3: System analysis

## 3.1 Requirements Specification

Our project is about the mobility of user from WLAN to 3G network. This process is known as vertical handoff in 4G, which is characterized by heterogeneity. The aims of our project are as the following:

1. Detecting vertical handoff in 4G, by defining suitable criteria.
2. Tracking the mobile user once the signal strength is below a certain threshold to execute the handoff process on time.
3. The tracking process will be conducted using the bounding box also which provides high accuracy and may be used for mobile network.
4. Using OMNeT++ to evaluate the performance of our system.

## 3.2 Requirements analysis

This part contains a detailed statistical analysis of the results to our survey. The results analysis includes answers from all respondents who took our survey in the 7-day period from Tuesday, November 11, 2014 to Monday, November 14, 2014 inclusive.

### 3.2.1 Results Analysis:

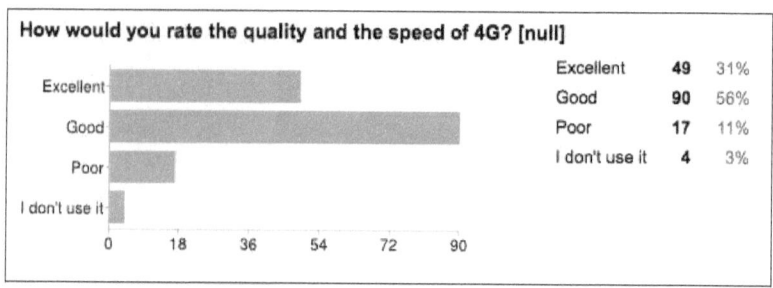

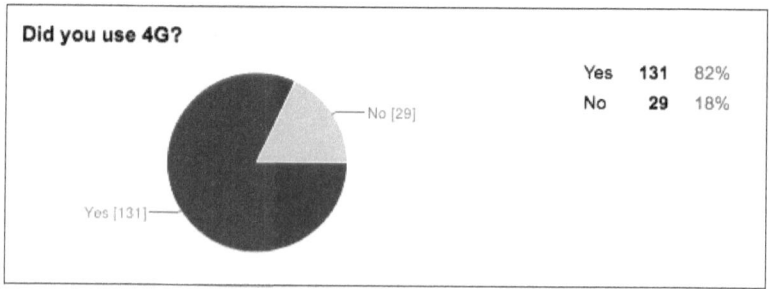

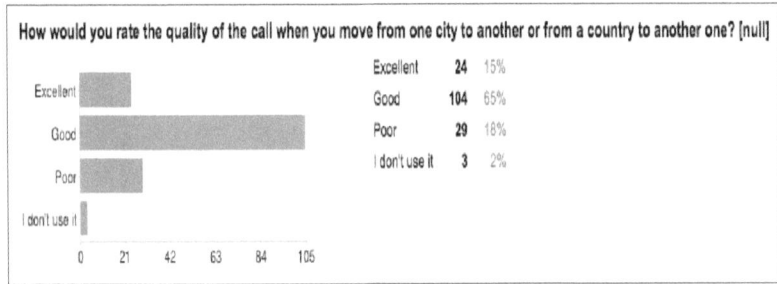

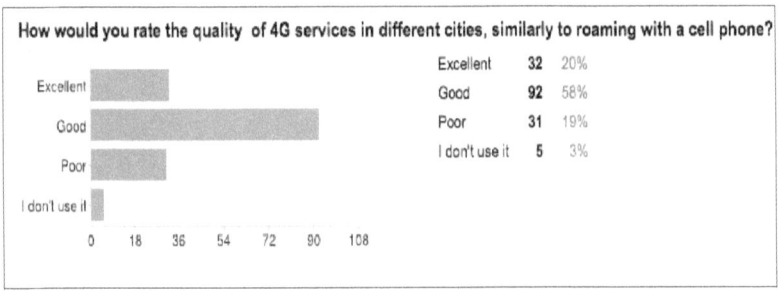

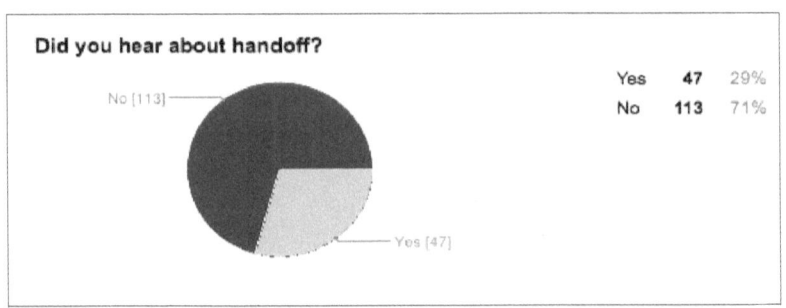

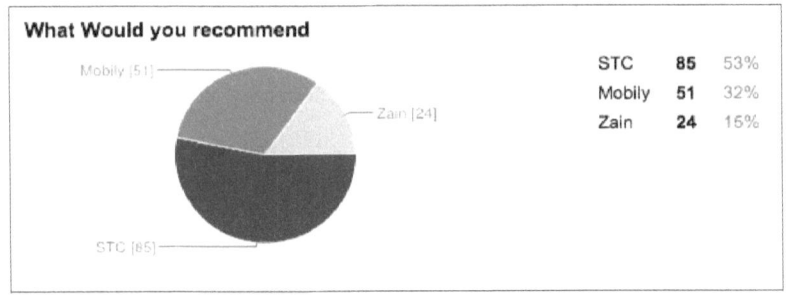

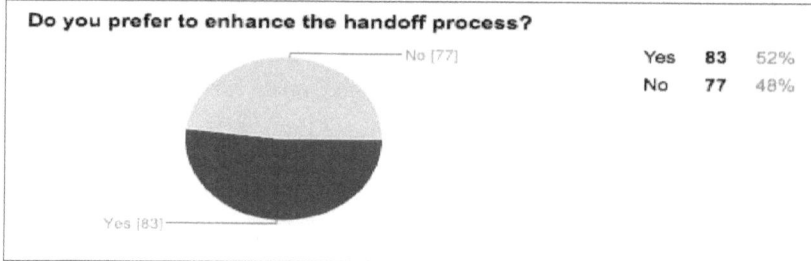

| | Yes | 83 | 52% |
|---|---|---|---|
| | No | 77 | 48% |

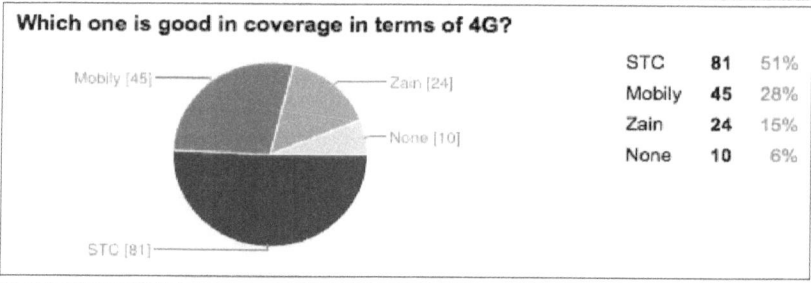

| | STC | 81 | 51% |
|---|---|---|---|
| | Mobily | 45 | 28% |
| | Zain | 24 | 15% |
| | None | 10 | 6% |

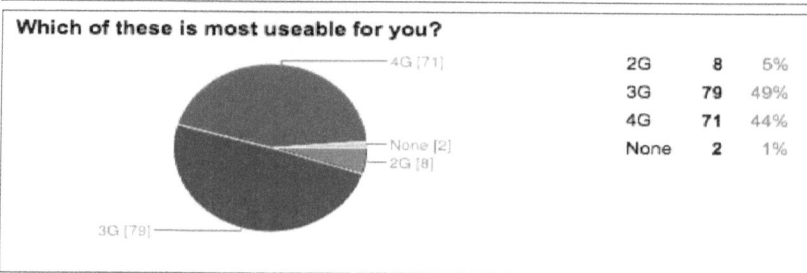

| | 2G | 8 | 5% |
|---|---|---|---|
| | 3G | 79 | 49% |
| | 4G | 71 | 44% |
| | None | 2 | 1% |

**In your opinion, what are the reasons of needing 4G?**

1. For more convenient usage of the network, save time, maybe it's usefull and needed more in the government organization more than personal usage.
2. Nowadays we need high speed of internet every second, if you just think about it! You will find that we cannot live a day with out being connected to Internet. So it is important to have a high speed internet such as 4G in our daily use.
3. High speed internet service due the need for internet in everything.
4. It's faster. Saves time when browsing.
5. Keep me connect wherever I go with fast and good data rete.

We conclude that the idea of our project is new: Most people are using 3G and enhancing the handoff process in 4G would attract people to use it.

**Chapter 4: System design**

## 4.1 System Architecture

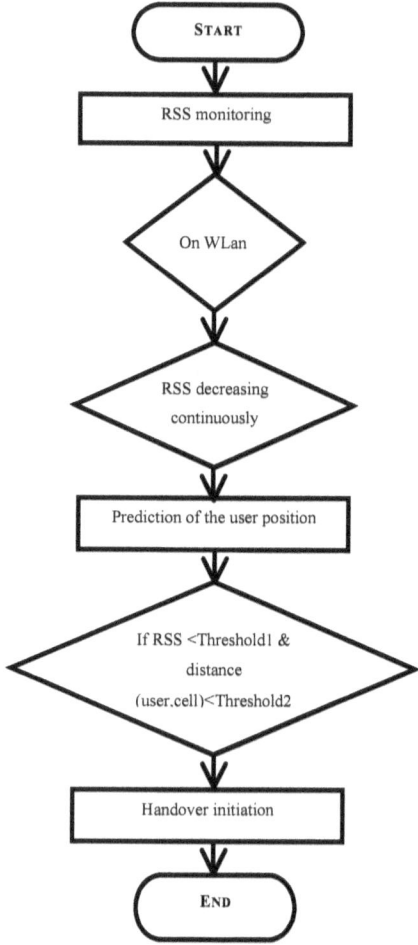

Figure 5: Flowchart for our system

- As you can see first we count the power of the signal by the RSS monitoring
- Then make sure that the user is moving from WLAN to 3G "because our method is only working in one way", if yes
- Test if the RSS is decreasing continuously!, if yes
- We should predicate the user position by the "bounding box algorithm"
- Check if the RSS is less than Threshold 1 & the distance (user, cell) is less than Threshold 2
- Starting the handover initiation.
- RSS= Constant Values
- Predicted Distance (user, cell)=$L_{BA}$

$$L_{BA} = \left[\tau^2 v^2 + d^2\left(p_f - 2 + 2\sqrt{1-p_f}\right)\right]^{\frac{1}{2}}.$$

- LBA is the shortest distance between the point at which handover is initiated
- 3G and WLAN boundary
- d is the side length of the WLAN cell (in meters, a WLAN cell is assumed to have a hexagonal shape in this study)
- $\epsilon \rightarrow$(in dB) is a zero-mean Gaussian random variable with a standard deviation that represents
- the statistical variation in RSS caused by shadowing.
- The distance LBA depends on the desired handover failure probability→ $p_f$
- the velocity of the mobile terminal → $v$
- the WLAN to 3G handover delay→ $\tau^2$

The AP will track the signal of the user movement and count the RSS, When the RSS is less than the threshold it doesn't necessarily means that the user is moving out to another cell (another AP) !! So we should predict the user movement by the "bounding box algorithm" (That will be used in our project)

- **RSS:**

    The received signal strength (RSS)-based approach to wireless localization offers the advantage of low cost and easy implement ability. (RSS) readily available and cost-effective method of location estimation, or localization, in wireless sensor networks (WSNs). However, RSS-derived distance estimates are known to be inaccurate, leading many researchers to conclude that RSS is an unreliable method for localization [12]

- **Bounding Box:**

    The bounding box algorithm is a computationally simple method of localizing nodes given their ranges to several beacons. See the next figure for an example. Essentially, each node assumes that it lies within the intersection of its beacons' bounding boxes. The bounding box for a beacon b is centered at the beacon position (xb, yb), and has height and width 2db, where db is the node's distance measurement to the beacon.[11]

    The intersection of the bounding boxes can be computed without use of floating point operations:

    $$[\max(x_i - d_i), \max(y_i - d_i)] \times [\min(x_i + d_i), \min(y_i + d_i)]$$

    $$i = 1 \ldots n$$

    The position of a node is then the center of this final bounding box, as shown in the next figure

    - Whitehouse analyzes a distributed version of this algorithm, showing that unfortunately this version is highly susceptible to noisy range estimates, especially small estimates which tend to propagate.
    - The accuracy of the bounding box approach is best when nodes' actual positions are closer to the center of their beacons. Simic and Sastry prove results about convergence, errors, and complexity.
    - In any event, bounding box works best when sensor nodes have extreme computational limitations, since other algorithms may simply be infeasible. Otherwise, more mathematically rigorous approaches such as gradient multilateration may be more appropriate.[11]

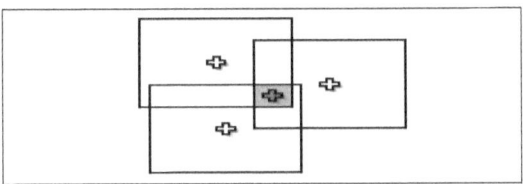

Figure 6: Bounding box algorithm

Above is an example of the intersection of bounding boxes. The center of the intersection is the position estimate for the unknown node. The size of the boxes is based on hop count radio range from the beacons to the unknown node.[11]

## 4.2 Comparison between our system and related works

Table 2: Comparison between our system and related works

| Group | Applicable Networking technologies | Input parameters | Handover target selection criteria | Complexity | Reliability |
|---|---|---|---|---|---|
| **RSS based VHD algorithms** | Usually between macro cellular and microcellular networks | RSS as the main input | The network candidate with the most stable RSS | Simple | Reduced reliability because of the fluctuation of RSS |
| **Bandwidth based VHD algorithms** | Between any two heterogeneous networks | Bandwidth combined with other parameters such as RSS | The network candidate with the highest bandwidth | Simple | Reduced reliability because of the changing available bandwidth |
| **Cost function based VHD algorithms** | Between any two heterogeneous networks | Various parameters such as cost, bandwidth and security | The network candidate with the highest overall performance | Complex | Reduced reliability because of its difficulty in measuring some parameters |
| **Combination algorithms** | Between any two heterogeneous networks | Different input parameters depending on different methods | The network candidate with the highest overall performance | Very complex | High reliability because of the training of the system |
| **Our system** | Between two heterogeneous networks | Cost, power, accuracy, mobile node and static node. | The network candidate with the lowest threshold | Simple | High reliability because of the static & mobile network appropriateness, accuracy |

## 4.3 User interface design

OMNeT++ is an object-oriented modular discrete event system simulation framework. It can be used in:

1. Modeling communication networks,
2. Protocol modeling, queuing networks,
3. Multiprocessors

Modeling and simulating a system via OMNeT++ can be conveniently mapped into entities communicating by exchanging messages.

OMNeT++ itself is not a simulator of anything concrete, but rather provides infrastructure and tools for writing simulations. One of the fundamental ingredients of this infrastructure is component architecture for simulation models.

### 4.3.1 Modeling Concepts:

An OMNeT++ model consists of modules that communicate with message passing. The active modules are termed simple modules; they are written in C++, using the simulation class library. Simple modules can be grouped into compound modules and so forth; the number of hierarchy levels is unlimited. The whole model, called network in OMNeT++, is itself a compound module.

Modules communicate through exchanging messages that may contain arbitrary data, in addition to usual attributes such as a timestamp. Simple modules typically send messages via gates, but it is also possible to send them directly to their destination modules.[20]

Gates are the input and output interfaces of modules: messages are sent through output gates and arrive through input gates. An input gate and output gate can be linked by a connection. Within a compound module, corresponding gates of two sub modules, or a gate of one sub module and a gate of the compound module can be connected.[20]

Connections spanning hierarchy levels are not permitted. Because of the hierarchical structure of the model, messages typically travel through a chain of connections, starting and arriving in simple modules

To perform the simulation for now all we need is modules and network [20]

- **Modules:**

    1. Node: is the devices in each network

    2. Point: is the access point in 3G network.

    3. Host: is the mobile user who will move from the networks to create the vertical handoff.

To connect the devices to each other we need to write a gate code to define that the module need a input and output in our simulation that would be a the connection. , And the code as following:

- **Gates:**

```
package final;

// auto generated module
simple Node
{
    @display("i=device/pocketpc");
    gates:
        input in[];
        output out[];
}
```

## 4.3.2 Interface simulation

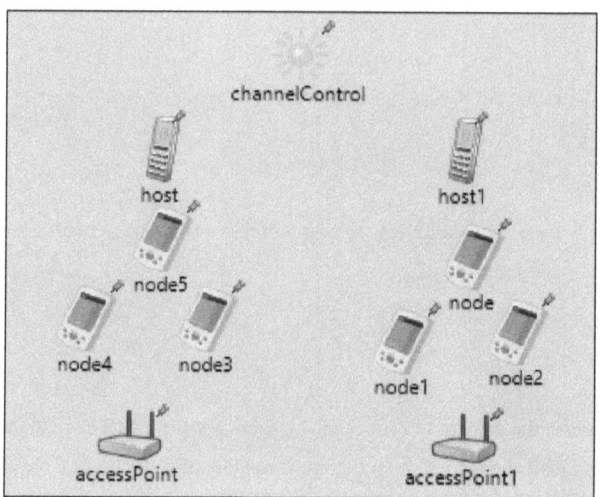

Figure 7: Interface Simulation

**Chapter 5: Implementation**

## 5.1 Implementation Requirement

### 5.1.1 Hardware:

To implement our project we used one device that has a processor 2.16 GHz, Memory 4 GB, Operating system Windows 64x Bit, and a modem to use the internet for software installation.

### 5.1.2 Software:

#### 5.1.2.1 OMNeT++:

As software we used OMNeT++ and the INet to have the mobility fetures imported from it, here is a small defintion for each framework, OMNeT++ framework is an extensible, modular, component-based C++ simulation library and framework, primarily used for building wired and wireless network simulations such as support for sensor networks, wireless adhoc networks, Internet protocols, performance modeling, photonic networks, etc. OMNeT++ offers an Eclipse-based IDE, a graphical runtime environment, and as some other tools. OMNeT++ provides component architecture for models. Components (modules) are programmed in C++, then assembled into larger components and models using a high-level language called NEtwork Description (NED). Reusability of models comes for free. All the modules and files of the OMNeT++ network simulator can be configured using GUI view or source view.[20]

#### 5.1.2.2 INET:

INET Framework contains IPv4, IPv6, TCP, SCTP, UDP protocol implementations, and several application models. The framework also includes an MPLS model with RSVP-TE and LDP signaling In the MPLS label switching routers (LSR) must agree on the meaning of the labels used to forward traffic between and through them. LDP (Label Distribution Protocol) is a new protocol that defines a set of procedures and messages by which one LSR (Label Switched Router) informs another of the label bindings it has made. RSVP-TE (traffic extension) protocol is an addition to the RSVP protocol (see TCP) with special extensions to allows it to set up optical paths in an agile optical network.) Link-layer models are PPP, Ethernet and 802.11. Static routing can be set up using network autoconfigurators, or one can use routing protocol implementations. The INET Framework supports wireless and mobile simulations as well. Support for mobility and wireless communication has been derived from the Mobility Framework.[16]

- o **INET uses the same concept as OMNeT++:**
  - Models consist of modules communicating by message passing (.msg)
  - Protocols are usually represented by simple modules in which external interfaces (gates/connectors and parameters) are described in a .ned file, and the implementation is prepared as a C++ class with the same name [16]

- o **INET modules:**
  - Protocol implementations (e.g. IPv4, IPv6, UDP, TCP, SCTP).
  - Auto configurators for whole network topologies (e.g. FlatNetworkConfigurator6).
  - Data storage and manipulation modules (e.g. InterfaceTable, RoutingTable6).
  - Managers for inter-module communication (e.g. NotificationBoard)
  - Radio channel managers (e.g. ChannelControl)
  - Implementations of host mobility (e.g. RandomWPMobility), etc.[16]

### 5.1.2.3 Modules and protocols:

Protocols are represented by simple modules. A simple module's external interface (gates [connectors] and parameters) is described in a NED file, and the implementation is contained in a C++ class with the same name. Some examples: TCP, IP. These modules can be freely combined to form hosts and other network devices with the NED language (no C++ code and no recompilation required). Various pre-assembled host, router, switch, access point, etc. models can be found in the Nodes/ subdirectory (for example: StandardHost, Router), but you can also create your own ones for tailored to your particular simulation scenarios.

Network interfaces (Ethernet, 802.11, etc) are usually compound modules (i.e. assembled from simple modules) themselves, and are being composed of a queue, a MAC, and possibly other simple modules. See EthernetInterface as an example.

Not all modules implement protocols though. There are modules which hold data (for example RoutingTable), facilitate communication of modules (NotificationBoard), perform autoconfiguration of a network (FlatNetworkConfigurator), move a mobile node around (for example ConstSpeedMobility), and perform housekeeping associated with radio channels in wireless simulations (ChannelControl). Protocol headers and packet formats are described in message definition files (msg files), which are translated into C++ classes by OMNeT++'s opp_msgc tool. The generated message classes subclass from OMNeT++'s cMessage class. [16]

### 5.1.2.4 Communication between protocol layers

In the INET Framework, when an upper-layer protocol wants to send a data packet over a lower-layer protocol, the upper-layer module just sends the message object representing the packet to the lower-layer module, which will in turn encapsulate it and send it. The reverse process takes place when a lower layer protocol receives a packet and sends it up after decapsulation.[16]

It is often necessary to convey extra information with the packet. For example, when an application-layer module wants to send data over TCP, some connection identifier needs to be specified for TCP. When TCP sends a segment over IP, IP will need a destination address and possibly other parameters like TTL. When IP sends a datagram to an Ethernet interface for transmission, a destination MAC address must be specified. This extra information is attached to the message object to as control info.[16]

Control info are small value objects, which are attached to packets (message objects) with its setControlInfo() member function. Control info only holds auxiliary information for the next protocol layer, and is not supposed to be sent over the network to other hosts and routers.[16]

### 5.1.2.5 Mobility Framework

This framework is intended to support wireless and mobile simulations within OMNeT++.The core framework implements the support for node mobility, dynamic connection management and a wireless channel model. Additionally the core framework provides basic modules that can be derived in order to implement own modules. With this concept a programmer can easily develop own protocol implementations for the Mobility Framework (MF) without having to deal with the necessary interface and interoperability stuff.[16]

o **Types of mobility:**

1. **Linear Mobility:** This is a linear mobility model with speed, angle and acceleration parameters, Angle only changes when the mobile node hits a wall: then it reflects off the wall at the same angle.z coordinate is constant movement is always parallel with X-Y plane[16]

2. **Circle Mobility:** Moves the node around a circle parallel to the X-Y plane with constant speed.The node bounces from the bounds of the constraint area. The circle is given by the cx, cy and r parameters, The initial position determined by the start Angleparameter. Theposition of the node is refreshed in updateInterval steps.[16]

3. **Rectangle Mobility:** Moves the node around the constraint area. configuration: speed, startPos, updateInterval[16]

4. **Random WP Mobility:** In the Random Waypoint mobility model the nodes move in line segments.For each line segment, a random destination position (distributed uniformly over the playground) and a random speed is chosen. You can define a speed as a variate from which a new value will be drawn for each line segment; it is customary to specify it as uniform(minSpeed, maxSpeed). When the node reache s the target position, it waits for the time waitTime which can also be defined as a variate.[16]

5. **Mass Mobility:** This is a random mobility model for a mobile host with a mass "An MH moves within the room according to the following pattern.[16]

6. **Chiang Mobility:** Chiang's random walk movement model.[16]

7. **Const Speed Mobility:** ConstSpeedMobility does not use one of the standard mobility approaches. the user can define a velocity for each Host and an update interval. If the velocity greater than zero (i.e. the Host is not stationary) the ConstSpeedMobility module calculates a random target position for the Host. Depending to the update interval and the velocity it calculates the number of steps to reach the destination and the step-size.[16]

8. **Ns2Mobility:** Nodes are moving according to the trace files used in NS2. [16]

## 5.2 Implementation Details:

Basically we created 2 networks each one has 3 nodes. The communication inside each network is managed through an access point. Then we added 2 hosts from the INet framework to move around the networks, and a channel control to control the movement, radio range, and the signals and we wrote the codes of sending and receiving in it. When you press play in the simulation interface the hosts will move and stop while they are sending and receiving the frames from/to the nodes in order to calculate the RSSI (Channel control) and then they move again and so on, till the RRSI reach a specific threshold that results to a handoff initiation process.

There are some common modules that appear in all (or many) host, router and device models.

1.

InterfaceTable. This module contains the table of network interfaces (eth0, wlan0, etc) in the host. This module does not send or receive messages: it is accessed by other modules using standard C++ member function calls. Other modules rely on the interface table submodule within the host to be called interfaceTable to be able to find it. (They obtain a cModule * pointer to it, then cast it to InterfaceTable * to be able to call its functions). Network interfaces get dynamically registered (added to the table) by simple modules implementing the network interface, for example EtherMAC.[16] [18] [19]

RoutingTable. This module contains the IP (v4) routing table, and heavily relies on InterfaceTable for its operation. This module is also accessed from other modules (typically IP) by calling the public member functions of its C++ class. There are member functions for querying, adding, deleting routes, and finding the best matching route for a given destination IP address. The routing table submodule within the host (router) must be called routingTable for other modules to find it.[16][18] [19]

NotificationBoard. This module makes it possible for several modules to communicate in a publish-subscribe manner. For example, the radio module (Ieee80211Radio) fires a "radio state changed" notification when

40

the state of the radio channel changes (from TRANSMIT to IDLE, for example), and it will be delivered to other modules that have previously subscribed to that notification category. The notification mechanism also works my C++ functions calls, no message sending is involved. The notification board submodule within the host (router) must be called notificationBoard for other modules to find it.[16] [18] [19]

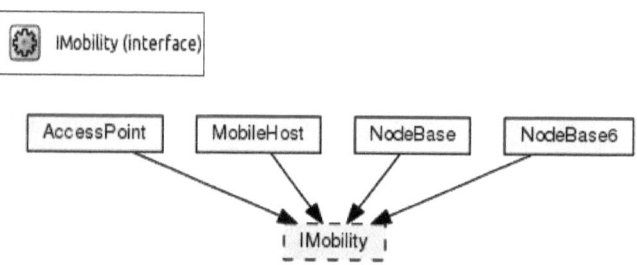

Figure 8: Mobility Architecture

Table 3: Description of Mobility Architecture

| Name | Type | Description |
| --- | --- | --- |
| AccessPoint | compound module | A generic access point supporting multiple wireless radios, and multiple Ethernet ports. The type of the Ethernet MAC, relay unit and wireless card can be specified as parameters. |
| MobileHost | compound module | A host for demonstrating mobility models only -- it contains no protocol layers at all. |
| NodeBase | compound module | Contains the common lower layers (linklayer and networklayer) of Router, StandardHost, WirelessHost etc. |
| NodeBase6 | compound module | BaseHost contains the common lower layers (linklayer and networklayer) of Router, StandardHost, WirelessHost etc |

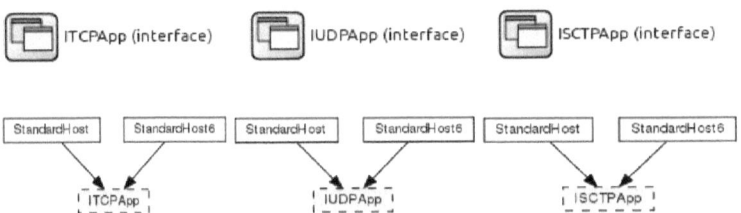

Table 4: Description of StandardHost

| Name | Type | Description |
|---|---|---|
| StandardHost | compound module | IPv4 host with SCTP, TCP, UDP layers and applications. IP forwarding is disabled by default (see IPForward). |
| StandardHost6 | compound module | IPv6 host with TCP, SCTP and UDP layers and applications. |

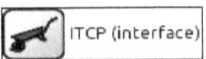 ITCP (interface)

Interface for TCP protocol implementations. All TCP implementations should implement this (i.e. declared as: TCP like ITCP) The existing implementations are these:TCP, TCP_NSC and TCP_lwIP.

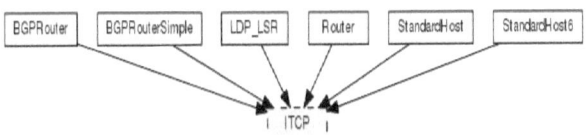

Figure 9: ITCP Architecture

Table 5: Description of ITCPQ Architecture

| Name | Type | Description |
|---|---|---|
| BGPRouter | compound module | Example IP router with BGPv4 and OSPFv4 support. |
| BGPRouterSimple | compound module | Example IPv4 router with BGPv4 support. |
| LDP_LSR | compound module | An LDP-capable router. |
| Router | compound module | IPv4 router that supports wireless, Ethernet, PPP and external interfaces. By default, no wireless and external interfaces are added; the number of Ethernet and PPP ports depends on the external connections |

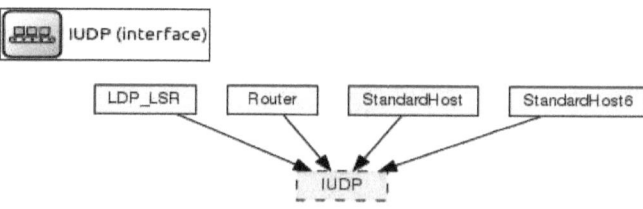

Figure 10: IUDP Architecture

Table 6: Description of IUDP Architecture

| Name | Type | Description |
|---|---|---|
| LDP_LSR | compound module | An LDP-capable router. |
| Router | compound module | IPv4 router that supports wireless, Ethernet, PPP and external interfaces. By default, no wireless and external interfaces are added; the number of Ethernet and PPP ports depends on the external connections. |

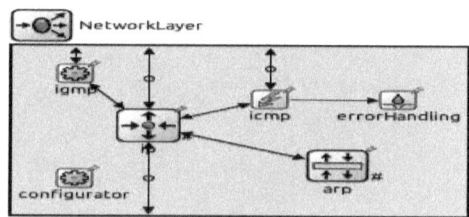

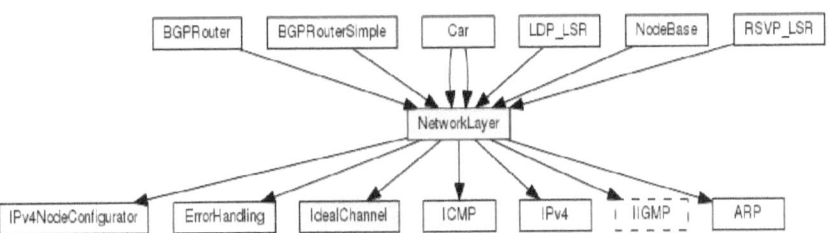

Figure 11: Network Layer Architecture

Table 7: Description of Network Layer Architecture

| Name | Type | Description |
| --- | --- | --- |
| BGPRouter | compound module | Example IP router with BGPv4 and OSPFv4 support. |
| BGPRouterSimple | compound module | Example IPv4 router with BGPv4 support. |
| LDP_LSR | compound module | An LDP-capable router. |
| NodeBase | compound module | Contains the common lower layers (linklayer and networklayer) of Router, StandardHost, WirelessHost etc. |
| RSVP_LSR | compound module | An RSVP-TE capable router. |

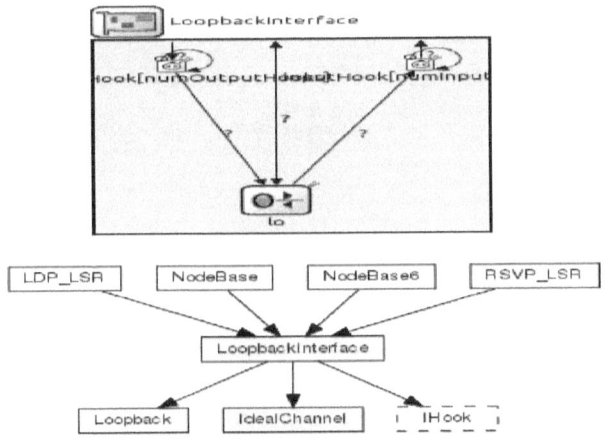

Figure 12: Loopback interface Architecture

Table 8: Description of Loopback interface Architecture

| Name | Type | Description |
|---|---|---|
| LDP_LSR | compound module | An LDP-capable router. |
| NodeBase | compound module | Contains the common lower layers (linklayer and networklayer) of Router, StandardHost, WirelessHost etc. |
| NodeBase6 | compound module | BaseHost contains the common lower layers (linklayer and networklayer) of Router, StandardHost, WirelessHost etc. |
| RSVP_LSR | compound module | An RSVP-TE capable router. |

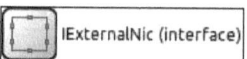

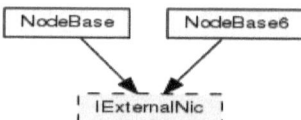

Figure 13: IExternalNic Architecture

Table 9: Description of Iexternalnic Architecture

| Name | Type | Description |
|---|---|---|
| NodeBase | compound module | Contains the common lower layers (linklayer and networklayer) of Router, StandardHost, WirelessHost etc. |
| NodeBase6 | compound module | BaseHost contains the common lower layers (linklayer and networklayer) of Router, StandardHost, WirelessHost etc. |

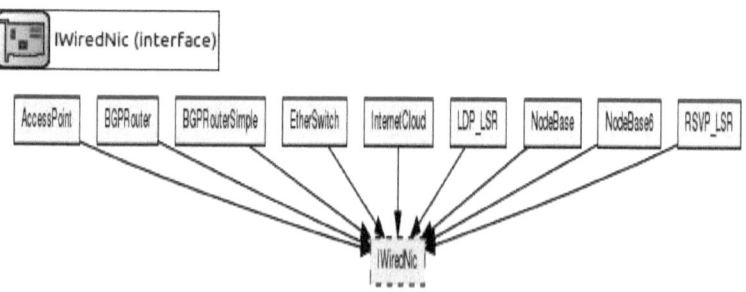

Figure 14: IWiredNic Architecture

- FlatNetworkConfigurator assigns IP addresses to hosts and routers, and sets up static routing.
- ScenarioManager makes simulations scriptable. Modules can be made to support scripting by implementing the IScriptable C++ interface.

2. ChannelControl

ChannelControl is required for wireless simulations. It keeps track of which nodes are within interference distance of other nodes

3. AccessPoint

A generic access point supporting multiple wireless radios, and multiple Ethernet ports. The type of the Ethernet MAC, relay unit and wireless card can be specified as parameters.

### 5.2.1 Simulation parameters:

We used these parameters to help us in our simulation and to make it more accurate:

Table 10: Simulation parameters

| Parameter | Value | Description |
| --- | --- | --- |
| Sim-time-limit | 60min | Simulation time limit |
| ChannelControl | 1 | The medium used to control wireless communications. |
| 3.CarrierFrequency | 2.4GHz | The frequency of signals. |
| 4.MobilityType | | Linear mobility |
| 5.Angle | 60deg | Mobility.angle of host 60 degrees |
| 6.Speed | 200mps | Host speed is 200 Meter Per Second |
| 7.Acceleration | -0.5 | Mobility acceleration -0.5 |
| 8.Interfaces | wlan0 | List of interfaces to send and receive data. |
| 9.MessageLength | 100B | Length Of Messages To Generate, In Bytes |
| 10.PathLossAlpha | 2 | The coefficient of packets loss in a wireless network. |

## 5.3 I/O screens:

### o Probability & RSSI

> The Value of RSSI is = -46.9471
>
> The probability is 0.566767 The Value of RSSI is = -48.1607
>
> The probability is 0.885215 The Value of RSSI is = -46.5
>
> The probability is 0.173103 The Value of RSSI is = -46.4446
>
> The probability is nan The Value of RSSI is = -46.6997
>
> The probability is 0.385219 The Value of RSSI is = -46.3958
>
> The probability is nan The Value of RSSI is = -47.5786
>
> The probability is 0.789318 The Value of RSSI is = -47.5786
>
> The probability is 0.789318 The Value of RSSI is = -47.1364
>
> The probability is 0.655771

### o AccessPoint

> Mynetwork.accessPoint.wlan[0].mac: ** Notification at T=20.011855442375 to Mynetwork.accessPoint.wlan[0].mac: RADIO-STATE TRANSMIT, channel #0, 2Mbps
>
> Mynetwork.accessPoint.wlan[0].mac: # state information: mode = DCF, state = WAITMULTICAST, backoff 0..1 = 0
>
> Mynetwork.accessPoint.wlan[0].mac: # backoffPeriod 0..1 = -1

> Mynetwork.accessPoint.wlan[0].mac: # retryCounter 0..1 = 0, radioState = 2, nav = 0, txop is 0
>
> Mynetwork.accessPoint.wlan[0].mac: #queue size 0..1 = 1, medium is ()busy, scheduled AIFS are 0(), scheduled backoff are 0
>
> Mynetwork.accessPoint.wlan[0].mac: # currentAC: 0, oldcurrentAC: 0
>
> Mynetwork.accessPoint.wlan[0].mac: # current transmission: 22433
>
> Mynetwork.accessPoint.wlan[0].mac: processing event in state machine Ieee80211Mac State Machine
>
> Mynetwork.accessPoint.wlan[0].mac: leaving handleWithFSM
>
> Mynetwork.accessPoint.wlan[0].mac: # state information: mode = DCF, state = WAITMULTICAST, backoff 0..1 = 0
>
> Mynetwork.accessPoint.wlan[0].mac: # backoffPeriod 0..1 = -1
>
> Mynetwork.accessPoint.wlan[0].mac: # retryCounter 0..1 = 0, radioState = 2, nav = 0, txop is 0
>
> Mynetwork.accessPoint.wlan[0].mac: #queue size 0..1 = 1, medium is busy, scheduled AIFS are 0()

- Node

> Mynetwork.node.wlan[0].mac: ** Notification at T=19.912672164392 to Mynetwork.node.wlan[0].mac: RADIO-STATE IDLE, channel #0, 2Mbps
>
> Mynetwork.node.wlan[0].mac: # state information: mode = DCF, state = IDLE, backoff 0..1 = 0
>
> Mynetwork.node.wlan[0].mac: # backoffPeriod 0..1 = 0
>
> Mynetwork.node.wlan[0].mac: # retryCounter 0..1 = 0, radioState = 0, nav = 0, txop is 0
>
> Mynetwork.node.wlan[0].mac: #queue size 0..1 = 0, medium is free, ()scheduled AIFS are 0(), scheduled backoff are 0

Mynetwork.node.wlan[0].mac: # currentAC: 0, oldcurrentAC: 0

Mynetwork.node.wlan[0].mac: # current transmission: none

Mynetwork.node.wlan[0].mac: processing event in state machine Ieee80211Mac State Machine

Mynetwork.node.wlan[0].mac: leaving handleWith
Mynetwork.node.wlan[0].mac: # state information: mode = DCF, state = IDLE, backoff 0..1 = 0

Mynetwork.node.wlan[0].mac: # backoffPeriod 0..1 = 0

- **Node1**

Mynetwork.node1.wlan[0].mac: ** Notification at T=20.011856001828 to Mynetwork.node1.wlan[0].mac: RADIO-STATE RECV, channel #0, 2Mbps

Mynetwork.node1.wlan[0].mac: # state information: mode = DCF, state = IDLE, backoff 0..1 = 0

Mynetwork.node1.wlan[0].mac: # backoffPeriod 0..1 = 0

Mynetwork.node1.wlan[0].mac: # retryCounter 0..1 = 0, radioState = 1, nav = 0, txop is 0

Mynetwork.node1.wlan[0].mac: #queue size 0..1 = 0, medium is busy, ()scheduled AIFS are 0(), scheduled backoff are 0

Mynetwork.node1.wlan[0].mac: # currentAC: 0, oldcurrentAC: 0

Mynetwork.node1.wlan[0].mac: # current transmission: none

Mynetwork.node1.wlan[0].mac: processing event in state machine Ieee80211Mac State Machine

Mynetwork.node1.wlan[0].mac: leaving handleWithFSM

- **Host**

  > Mynetwork.host.wlan[0].mac: ** Notification at T=20.011855879187 to Mynetwork.host.wlan[0].mac: RADIO-STATE RECV, channel #0, 2Mbps
  >
  > Mynetwork.host.wlan[0].mac: # state information: mode = DCF, state = IDLE, backoff 0..1 = 0
  >
  > Mynetwork.host.wlan[0].mac: # backoffPeriod 0..1 = 0
  >
  > Mynetwork.host.wlan[0].mac: # retryCounter 0..1 = 0, radioState = 1, nav = 0, txop is 0
  >
  > Mynetwork.host.wlan[0].mac: #queue size 0..1 = 0, medium is busy, ()scheduled AIFS are 0(), scheduled backoff are 0
  >
  > Mynetwork.host.wlan[0].mac: # currentAC: 0, oldcurrentAC: 0
  >
  > Mynetwork.host.wlan[0].mac: # current transmission: none
  >
  > Mynetwork.host.wlan[0].mac: processing event in state machine Ieee80211Mac State Machine
  >
  > Mynetwork.host.wlan[0].mac: leaving handleWithFSM
  >
  > # state information: mode = DCF, state = Mynetwork.host.wlan[0].mac: IDLE, backoff 0..1 = 0

- **Host1**

  > Mynetwork.host1.wlan[0].mac: ** Notification at T=20.011855572164 to Mynetwork.host1.wlan[0].mac: RADIO-STATE RECV, channel #0, 2Mbps
  >
  > Mynetwork.host1.wlan[0].mac: # state information: mode = DCF, state = IDLE, backoff 0..1 = 0
  >
  > Mynetwork.host1.wlan[0].mac: # backoffPeriod 0..1 = 0
  >
  > Mynetwork.host1.wlan[0].mac: # retryCounter 0..1 = 0, radioState = 1, nav = 0, txop is 0

Mynetwork.host1.wlan[0].mac: #queue size 0..1 = 0, medium is busy, ()scheduled AIFS are 0(), scheduled backoff are 0

Mynetwork.host1.wlan[0].mac: # currentAC: 0, oldcurrentAC: 0

Mynetwork.host1.wlan[0].mac: # current transmission: none

Mynetwork.host1.wlan[0].mac: processing event in state machine Ieee80211Mac State Machine

Mynetwork.host1.wlan[0].mac: leaving handleWithFSM

# state information: mode = DCF, state = Mynetwork.host1.wlan[0].mac: IDLE, backoff 0..1 = 0

Mynetwork.host1.wlan[0].mac: # backoffPeriod 0..1 = 0

Mynetwork.host1.wlan[0].mac: # retryCounter 0..1 = 0, radioState = 1, nav = 0, txop is 0

# Chapter 6: Testing

## 6.1 Test plan:

To make sure that our method is working well; we are going to test host moving around and see the RSSI values.

## 6.2 Test case:

Table 11: Test case

| Case 1 | Refer to Fig #16 The values of the RSSI while the host moving around |
|--------|---------------------------------------------------------------------|
| Case 2 | Refer to fig #17 Evaluating the impact of the speed |
| Case 3 | Refer to fig #18 Handoff probability |

## Test result:

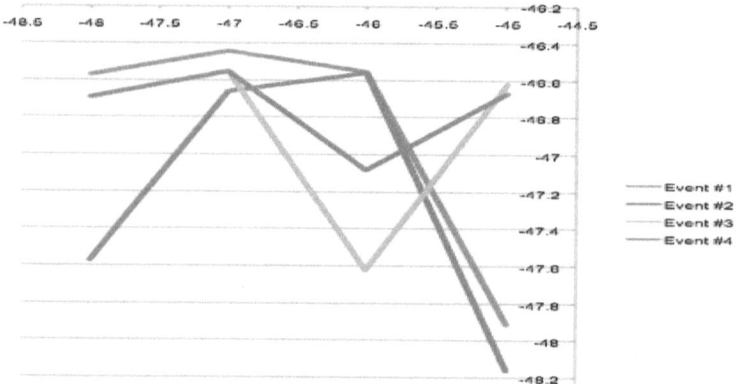

Figure 15: RSSI average on 200 speed result

As you can see above when the smulation runs it keeps gnetarion handoff in diffrenet time, Figure 15 show you 4 genration handoff.

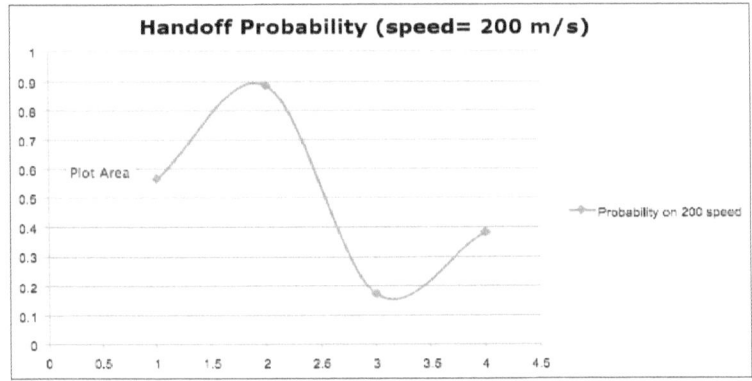

Figure 16: Probability on 200 speed result

Figure 16 shows the handoff probability for a host that moves with 200mps as a speed. The X axis shows the simulation time. It is clear that the handoff probability incresses once the host goes far away from the first access point and then decreases once the host become near the second access point.

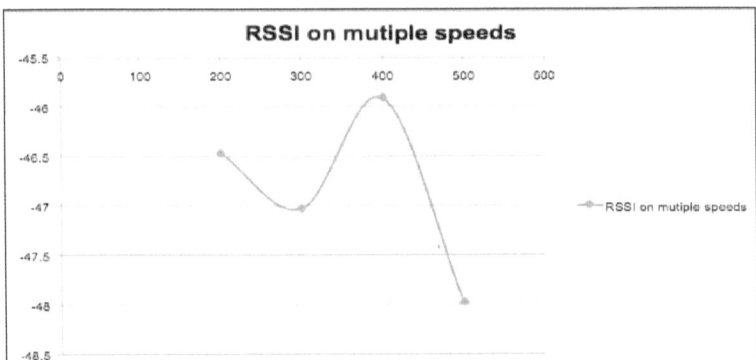

Figure 17: RSSI on multiple speeds result

Figure 17 show the RSSI in multiple speed and occurrence of the handoff. The Y axis shows the several values where the handoff initiate and you can notice that the probability must be less than 1 for the handoff to starts, the X axis show how the RSSI decrease according the speed of the host once its value is greater than 400 M/S.

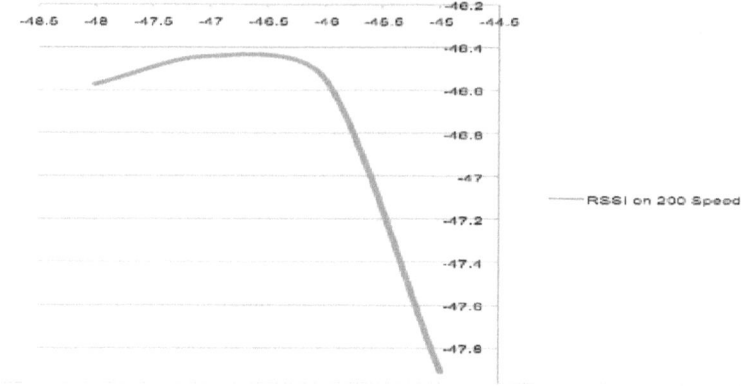

Figure 18: RSSI on 200 speeds result

Figure 18 show clearly how the handoff will be on 200mps , conculading that Once the RSSI < a certain speed the handoff will start.

# Chapter 7: Conclusion

## 7.1 Evaluation:

We are very proud that the method we used to calculate the RSSI came up with a result as we expected:

1. The signal strength decreases according to the simulation time.

2. The signal strength decreases starting from a certain threshold.

3. The RSSI decrease according the speed of the host once its value is greater than 400 M/S.

4. Then handoff probability increases according to the time of simulation and then decreases once the host is near to the second access point.

## 7.2 Future work:

Unfortunately we didn't have enough time to implement the Bounding Box algorithm witch is a localization method that would decrease the consumption power used for handoff that's why we assumed that the position of the hosts are known.

## References:

1. F. Williams, Ericsson, "Fourth generation mobile," in ACTS Mobile Summit99, Sorrento, Italy, June 1999.

2. H. Huomo, Nokia, "Fourth generation mobile," in ACTS Mobile Summit99, Sorrento, Italy, June 1999.

3. Jun-Zhao Sun, Jaakko Sauvola, and Douglas Howie, "Features in Future: 4G Visions From a Technical Perspective," in IEEE, 2001.

4. Mishra, Ajay K. "Fundamentals of Cellular Network Planning and Optimization, 2G/2.5G/3G…Evolution of 4G", John Wiley and Sons, 2004.

5. Pereira, Vasco & Sousa, Tiago. "Evolution of Mobile Communications: from 1G to 4G", Department of Informatics Engineering of the University of Coimbra, 2004.

6. "An Overview of Vertical Handoff Decision Making Algorithms" - A. Bhuvaneswari, Dr. E. George Dharma Prakash Raj- I. J. Computer Network and Information Security, (2012).

7. "A Comparison Of Different Vertical Handoff Algorithms Between Wlan And Cellular Networks"- Elaheh Arabmakki-IEEE Communications Magazine(2008).

8. "A New Fuzzy Simulation Model for Vertical Handoff in Heterogeneous Networks"-Harvinder Gill, Silki Baghla, SSRG International Journal of Electronics and Communication Engineering (SSRG-IJECE) – volume1 issue6 (2006).

9. "Vertical Handover between Wi-Fi and WiMAX"- Ankur Saini, Preeti Bhalla,- International Journal of Advanced Research in Computer Science and Software Engineering (2013).

10. "Vertical handover criteria and algorithm in IEEE 802.11 and 802.16 hybrid networks"- Z. Dai, R. Fracchiaa, J. Gosteaub, Motorola Labs Paris-Parc les Algorithmes de Saint-Aubin, 91193 Gif-sur-Yvette, France(2007).

11. "Localization in Sensor Networks"- Jonathan Bachrach and Christopher Taylor, Computer Science and Artificial Intelligence Laboratory Massachusetts Institute of Technology (2005).

12. "Design and Implementation of Reliable Localization Algorithms using

Received Signal Strength"- Jeffrey Vander Stoep, University of Washington (2009).

13. "Effect of varied wireless standards and properties towards wireless network bandwidth"- Wooi King Soo, Teck-Chaw Ling and Keat-Keong Phang, Proceedings of the Asia-Pacific Advanced Network (2013).

14." I-TCP: Indirect TCP for Mobile Hosts" Department of Computer Science Rutgers University, Piscataway Ajay Bakre NJ 08855. DCS-TR-314 October, (1994).

15. "Two-level event brokering architecture for information dissemination in vehicular networks" Tina Devkota, Submitted to the Faculty of the Graduate School of Vanderbilt University in partial fulfillment of the requirements for the degree of master of science(May 2009).

16. "A Mobility Framework for OMNeT++ User Manual"- Marc Lobbers " Daniel Willkomm - January 12, (2007).

17. "Simulation f Mobile IPv6 Using OMNeT++ Simulator" Dr. Emad H. Al-Hemiary2-(2011).

18. "INET Framework for OMNeT++Manua"- Generated on June 21, (2012).

19. http://klub.com.pl/numbat/neddoc/inet-architecture.html

20. OMNeT++ User Guide Version 4.6

## Appendices:
- **Omnet.ini**

```
[General]
network = Mynetwork
*.numOfHosts = 5
sim-time-limit = 60min
cmdenv-express-mode = true
#debug-on-errors = true
tkenv-plugin-path = ../../../etc/plugins
**.constraintAreaMinX = 0m
**.constraintAreaMinY = 0m
**.constraintAreaMinZ = 0m
**.constraintAreaMaxX = 600m
**.constraintAreaMaxY = 400m
**.constraintAreaMaxZ = 0m
**.debug = true
**.coreDebug = false
**.host*.**.channelNumber = 0
# channel physical parameters
*.channelControl.carrierFrequency = 2.4GHz
*.channelControl.pMax = 2.0mW
*.channelControl.sat = -110dBm
*.channelControl.alpha = 2
```

```
*.channelControl.numChannels = 1

# mobility

network = MobileNetwork

**.host*.mobilityType = "LinearMobility"

**.host*.mobility.initFromDisplayString = false

**.host*.mobility.speed = 200mps

**.host*.mobility.angle = 60deg  # degrees

**.host*.mobility.acceleration = -0.5

**.host*.ac_wlan.interfaces = "wlan0"

# UDPBasicApp / UDPSink

**.numUdpApps = 1

**.udpApp[0].typename = "UDPBasicApp"

**.udpApp[0].destAddresses = "host[0]"

**.udpApp[0].localPort = 9001

**.udpApp[0].destPort = 9001

**.udpApp[0].messageLength = 100B

**.udpApp[0].startTime = uniform(10s, 30s)

**.udpApp[0].sendInterval = uniform(10s, 30s)

# nic settings

**.wlan[*].bitrate = 2Mbps

**.wlan[*].mgmt.frameCapacity = 10

**.wlan[*].mac.address = "auto"

**.wlan[*].mac.maxQueueSize = 14
```

```
**.wlan[*].mac.rtsThresholdBytes = 3000B

**.wlan[*].mac.retryLimit = 7

**.wlan[*].mac.cwMinData = 7

**.wlan[*].mac.cwMinBroadcast = 31

**.wlan[*].radio.transmitterPower = 2mW

**.wlan[*].radio.thermalNoise = -110dBm

**.wlan[*].radio.sensitivity = -85dBm

**.wlan[*].radio.pathLossAlpha = 2

**.wlan[*].radio.snirThreshold = 4dB

**.udpapp.*.vector-recording = true

**.vector-recording = true
```

- ChannelControl

```
  const RadioRefVector& neighbors = getNeighbors(srcRadio);
  int n = neighbors.size();
  int channel = airFrame->getChannelNumber();
  for (int i=0; i<n; i++)
  {
     RadioRef r = neighbors[i];
     if (!r->isActive)
     {
       coreEV << "skipping disabled radio interface \n";
       continue;
     }
     if (r->channel == channel)
     {
       coreEV << "sending message to radio listening on the same channel\n";
       // account for propagation delay, based on distance in meters
       // Over 300m, dt=1us=10 bit times @ 10Mbps
       simtime_t delay = srcRadio->pos.distance(r->pos) / SPEED_OF_LIGHT;
                    check_and_cast<cSimpleModule*>(srcRadio->radioModule)-
>sendDirect(airFrame->dup(), delay, airFrame->getDuration(), r->radioInGate);
       double rssi = log (srcRadio->pos.distance(r->pos)* 24)- 55;
       EV<<"The Value of RSSI is = "<<rssi<<endl;

       if( rssi < - 46.629)
       {
          bubble("handoff initiation");

       }
        double Probability = acos ((srcRadio->pos.distance(r->pos) / 200 )) / atan (
maxInterferenceDistance  / (2* srcRadio->pos.distance(r->pos)) );
       ev<<"The probability is "<<Probability;
    }

    else
       coreEV << "skipping radio listening on a different channel\n";
  }
  // register transmission
  addOngoingTransmission(srcRadio, airFrame);
}
```

# I want morebooks!

Buy your books fast and straightforward online - at one of the world's fastest growing online book stores! Environmentally sound due to Print-on-Demand technologies.

## Buy your books online at
## www.get-morebooks.com

Kaufen Sie Ihre Bücher schnell und unkompliziert online – auf einer der am schnellsten wachsenden Buchhandelsplattformen weltweit!
Dank Print-On-Demand umwelt- und ressourcenschonend produziert.

## Bücher schneller online kaufen
## www.morebooks.de

OmniScriptum Marketing DEU GmbH
Heinrich-Böcking-Str. 6-8
D - 66121 Saarbrücken
Telefax: +49 681 93 81 567-9

info@omniscriptum.com
www.omniscriptum.com

www.ingramcontent.com/pod-product-compliance
Lightning Source LLC
Chambersburg PA
CBHW031539210526

45464CB00003B/1072